Alligators In SPACE

Written and Illustrated by Jenny Dearinger

Copyright 2024 by Jenny Dearinger and Everfield Press
All rights reserved. No part of this publication including pictures may be reproduced, stored in a retrieval system, of transmitted in any form or by any means, electronic, mechanical, recording or otherwise, without prior written permission of the author. This book is for entertainment purposes only. The views expressed are the author's alone.

SUN

Whew! The sun is way too hot!
The only alien who could live here would be
A super robot,
Let's leave before we melt into a sunspot.

FACTS

The Sun is at the center of our Solar System

The Sun is a ball of gas that's on fire.

The Sun is so big, over 1 million Earths could fit into it.

It takes the light from the Sun 8 minutes and 20 seconds to reach the Earth.

MERCURY

Mercury is almost as hot as
The sun.
These aliens must like their
Food extra well done.
If that volcano goes off,
Please call someone.

FUN FACTS
- Mercury is the smallest planet.
- It is very hot. There are no active volcanoes on Mercury.
- One year on Mercury is only 88 Earth days.

MARS

Do little green men live on Mars?
Do they drive around in blue and orange cars?
Do they scream and sing along with Mars pop-stars?

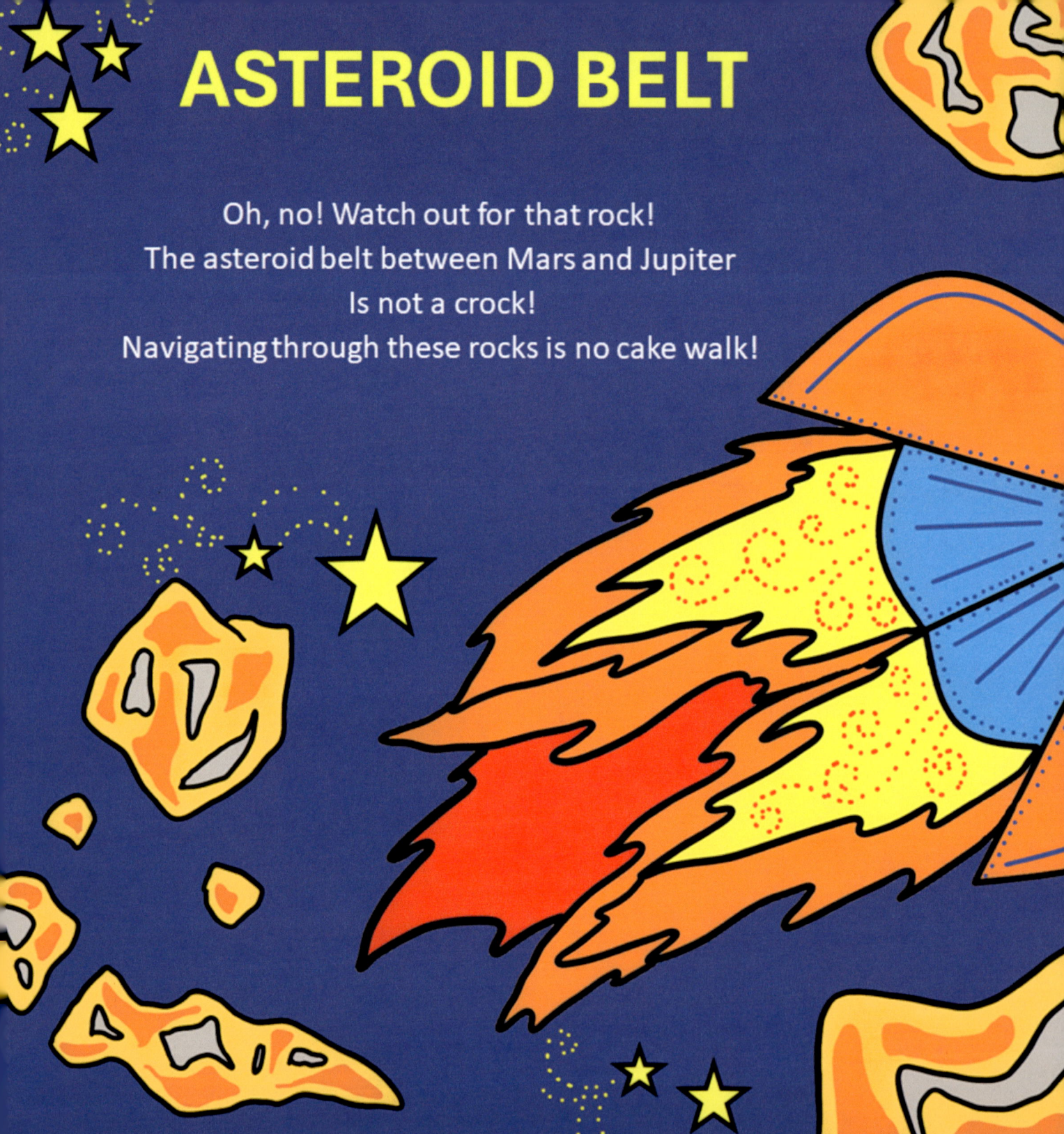

ASTEROID BELT

Oh, no! Watch out for that rock!
The asteroid belt between Mars and Jupiter
Is not a crock!
Navigating through these rocks is no cake walk!

FUN FACTS
- The Asteroid Belt is made up of almost 2 million rocks that never formed into a planet.
- The largest rock in the Asteroid Belt is Ceres. In 2006, Ceres was designated a dwarf planet like Pluto.

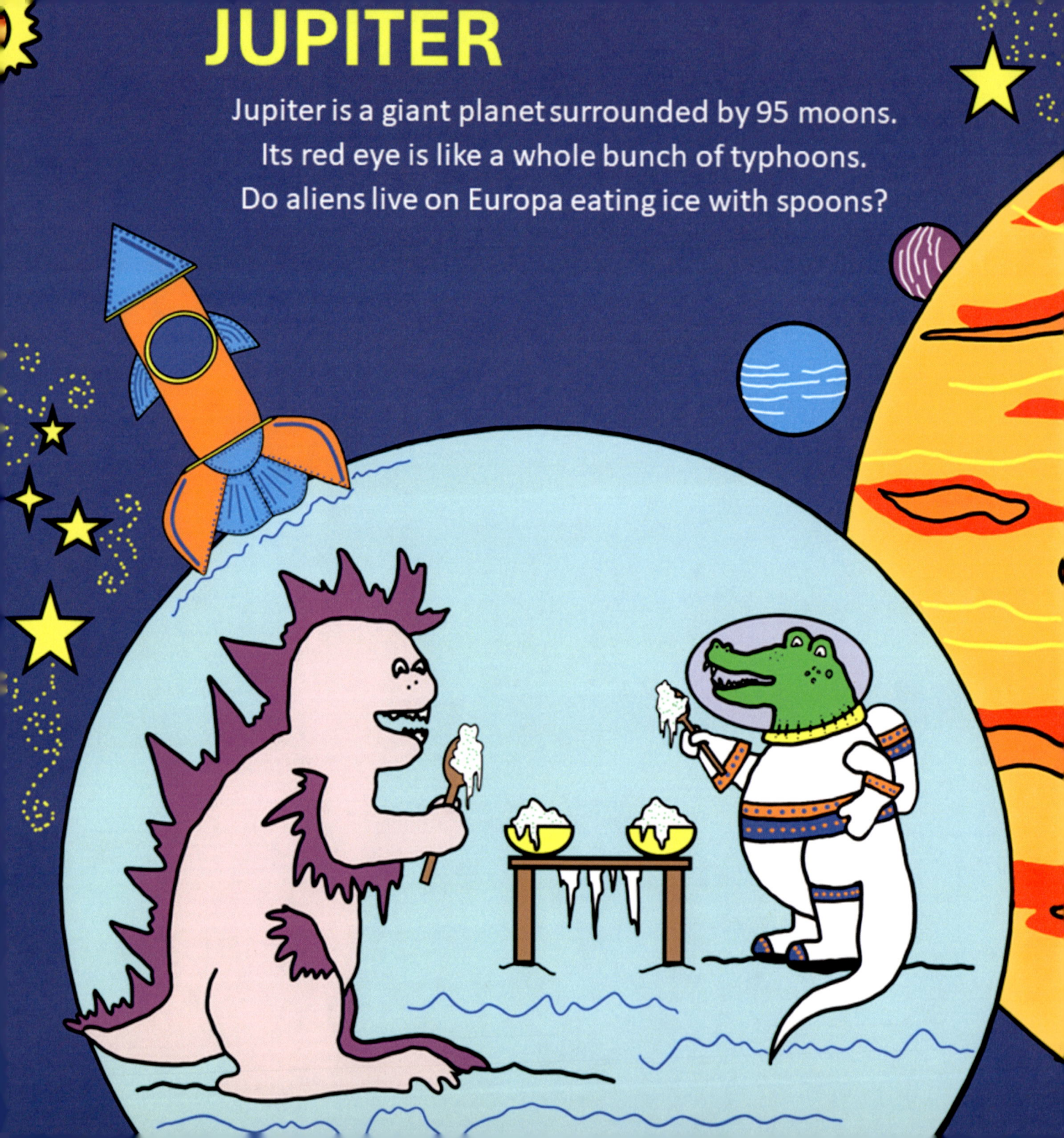

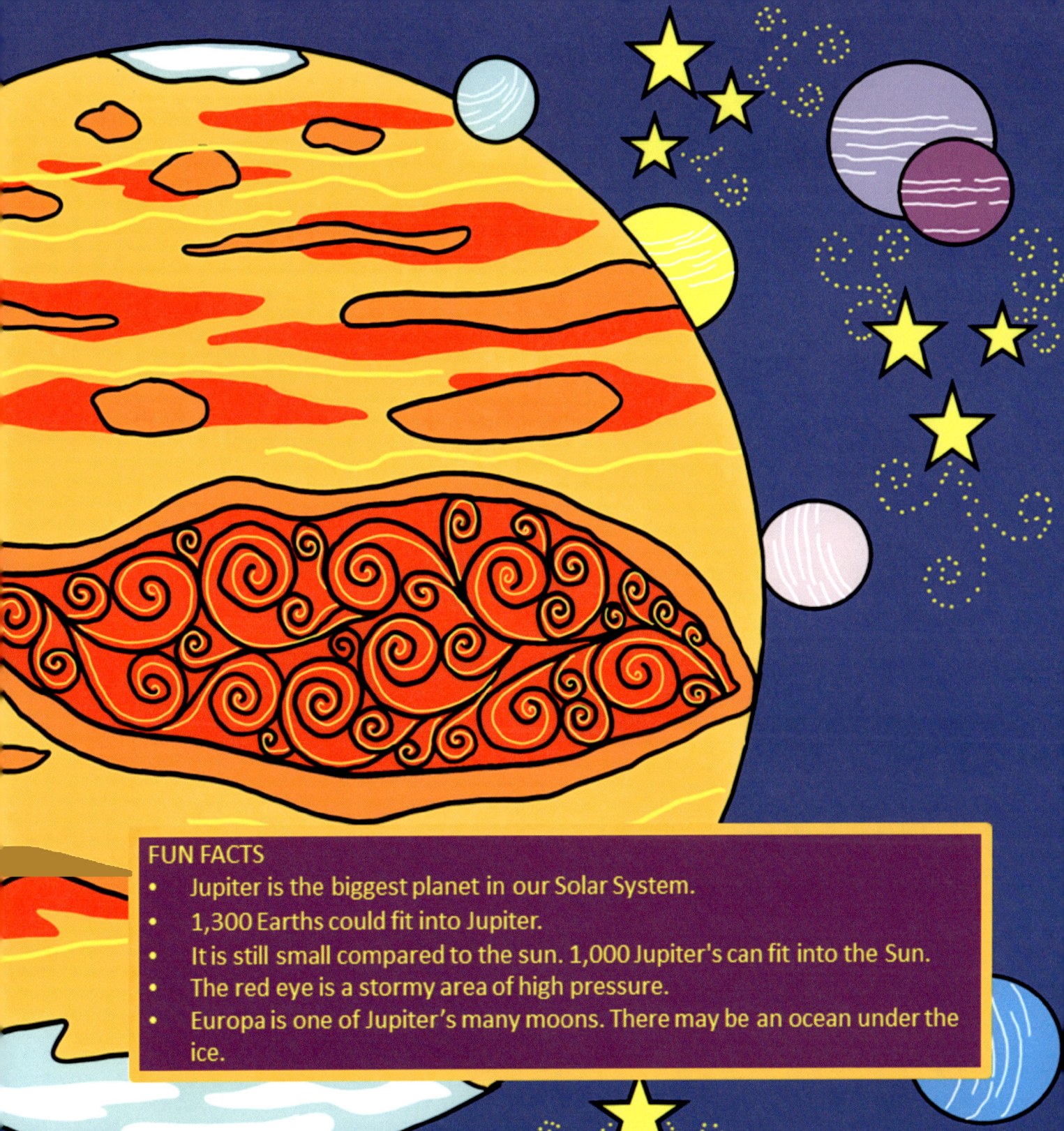

FUN FACTS
- Jupiter is the biggest planet in our Solar System.
- 1,300 Earths could fit into Jupiter.
- It is still small compared to the sun. 1,000 Jupiter's can fit into the Sun.
- The red eye is a stormy area of high pressure.
- Europa is one of Jupiter's many moons. There may be an ocean under the ice.

SATURN

Ooh! How pretty!
Look past the rings.
Do you see an inner City?
Do the aliens there have a welcoming
Committee?

FUN FACTS
- Saturn is called 'the Ringed Planet.'
- Saturn has 7-8 main rings around it made up of lots of smaller rings and gaps.
- The rings are made up of trapped asteroids, shattered moons, and clumps of icy dust.

URANUS

Let's visit aliens on Uranus.
Will they throw a parade when they see us?
We don't want them to go out of their way
And make a big fuss.

FUN FACTS
- Uranus is an icy blue planet.
- It is made up of mostly methane, ammonia, hydrogen, and helium gases.
- The cool thing about Uranus is that its rings run up and down because its equator runs north to south instead of east to west like Earth's.

NEPTUNE

Is Neptune a giant ocean full of fish?
Do the aliens there cook a tasty fish dish?
When they see Earth twinkle like a star do
They make a wish?

FUN FACTS
- Like Uranus, Neptune is an icy giant.
- Remember that it only takes Mercury 88 days to make a year? It takes Neptune 165 earth years to go around the sun one time!

PLUTO

FUN FACTS
- In 2006, Pluto was downgraded from a planet to a "dwarf" planet. Pluto is 40 times further from the sun than Earth.
- The Earth's moon is 2,100 miles across. Pluto is only 1,477 miles across which makes it smaller than the Moon.

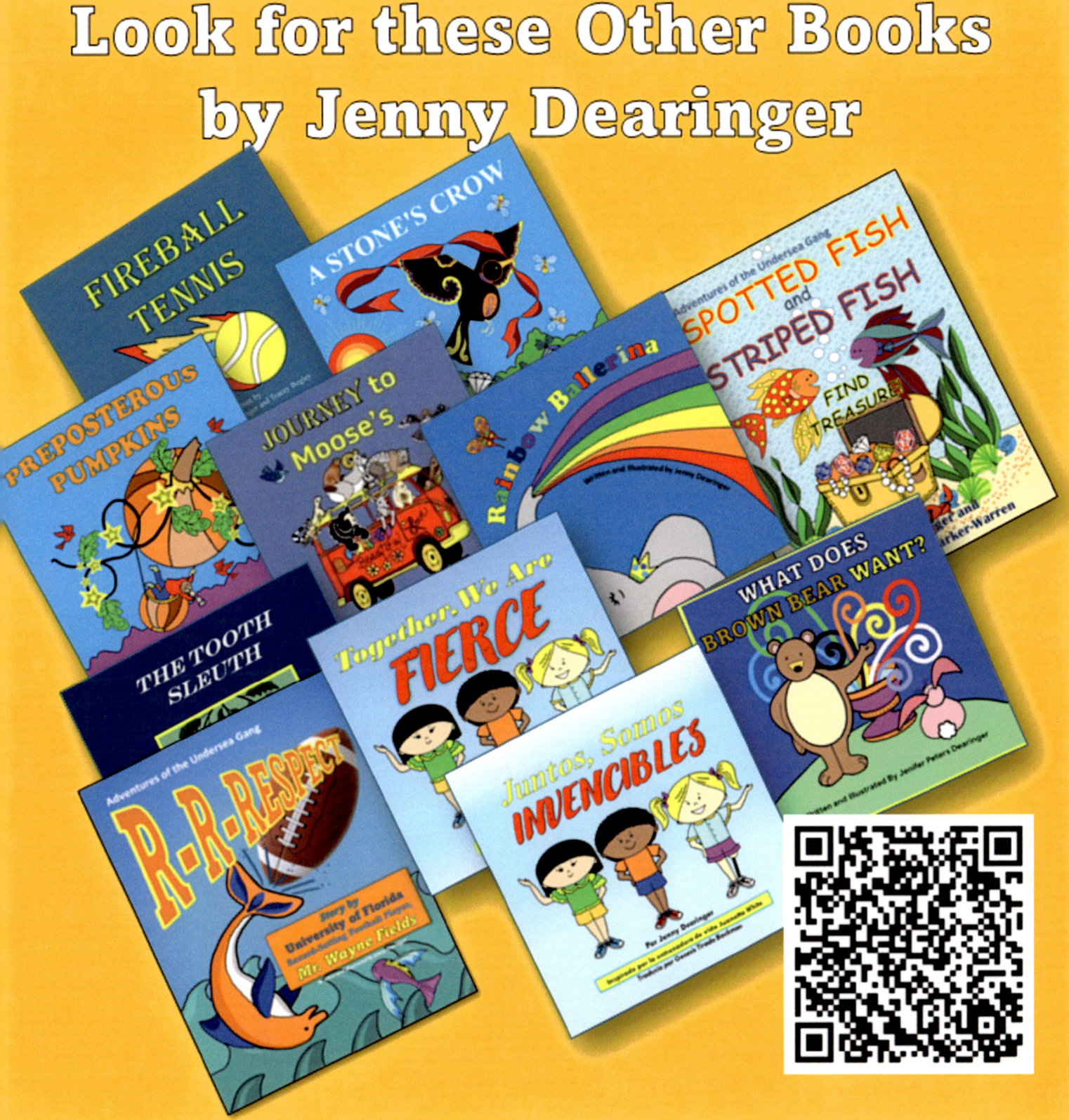

www.ingramcontent.com/pod-product-compliance
Lightning Source LLC
Chambersburg PA
CBRC091726070526
44586CB00008B/89